Textile Progress

2008   Vol 40   No 2

Fibre materials for advanced technical textiles

T. Matsuo

The Textile Institute

Taylor & Francis

Taylor & Francis

SUBSCRIPTION INFORMATION

Textile Progress (USPS Permit Number pending), Print ISSN 0040-5167, Online ISSN 1754-2278, Volume 40, 2008.

Textile Progress (www.informaworld.com/textileprogress) is a peer-reviewed journal published quarterly in March, June, September and December by Taylor & Francis, 4 Park Square, Milton Park, Abingdon, Oxon, OX14 4RN, UK on behalf of The Textile Institute.

Institutional Subscription Rate (print and online): $304/£160/€243
Institutional Subscription Rate (online-only): $288/£152/€230 (plus tax where applicable)
Personal Subscription Rate (print only): $112/£58/€90

All current institutional subscriptions include online access for any number of concurrent users across a local area network to the currently available backfile and articles posted online ahead of publication.

Ordering Information: Please contact your local Customer Service Department to take out a subscription to the Journal: **India**: Universal Subscription Agency Pvt. Ltd, 101–102 Community Centre, Malviya Nagar Extn, Post Bag No. 8, Saket, New Delhi 110017. **Japan**: Kinokuniya Company Ltd, Journal Department, PO Box 55, Chitose, Tokyo 156. **USA, Canada and Mexico**: Taylor & Francis, 325 Chestnut Street, 8th Floor, Philadelphia, PA 19106, USA. Tel: +1 800 354 1420 or +1 215 625 8900; fax: +1 215 625 8914, email: customerservice@taylorandfrancis.com. **UK and all other territories**: T&F Customer Services, Informa Plc., Sheepen Place, Colchester, Essex, CO3 3LP, UK. Tel: +44 (0)20 7017 5544; fax: +44 (0)20 7017 5198, email: tf.enquiries@tfinforma.com.

Dollar rates apply to all subscribers outside Europe. Euro rates apply to all subscribers in Europe, except the UK and the Republic of Ireland where the pound sterling price applies. All subscriptions are payable in advance and all rates include postage. Journals are sent by air to the USA, Canada, Mexico, India, Japan and Australasia. Subscriptions are entered on an annual basis, i.e. January to December. Payment may be made by sterling cheque, dollar cheque, euro cheque, international money order, National Giro or credit cards (Amex, Visa and Mastercard).

Back Issues: Taylor & Francis retains a three year back issue stock of journals. Older volumes are held by our official stockists to whom all orders and enquiries should be addressed:
Periodicals Service Company, 11 Main Street, Germantown, NY 12526, USA. Tel: +1 518 537 4700; fax: +1 518 537 5899; email: psc@periodicals.com.

Periodical postage paid at Jamaica, NY 11431, by US Mailing Agent Air Business Ltd, c/o Worldnet Shipping USA Inc., 149-35 177th Street, Jamaica, New York, NY 11434.

For more information on Taylor & Francis' journal publishing program, please visit our website: www.informaworld.com/journals.

CONTENTS

1. **Introduction**   87
    1.1. The content and objective of this article   87
    1.2. What is advanced technical textiles?   87
    1.3. Why are fibre materials used for technical products?   88
    1.4. General scope on fibre materials for advanced technical textiles   88
    1.5. General scope on technical textile products as the application items
          of fibre materials   89

2. **Conventional fibres**   90
    2.1. General scope on conventional fibres   90
    2.2. PET (polyethylene terephthalate) fibre   91
    2.3. Nylon fibre   91
    2.4. Polypropylene fibre   92

3. **High mechanical performance fibres**   92
    3.1. General scope on high mechanical performance fibres   92
    3.2. Carbon fibre   93
    3.3. Glass fibre   94
    3.4. P-aramid fibre   94
    3.5. UHMW-PE (ultra-high molecular weight polyethylene) fibre   96
    3.6. Wholly aromatic polyester fibre   96
    3.7. Other mechanical high performance fibres   96
          3.7.1   *Poly-p-phenylene benzobisoxazole fibre*   96
          3.7.2   *PIPD fibre*   97
          3.7.3   *Polyketone fibre*   97
          3.7.4   *Basalt fibre*   97
          3.7.5   *SiC group fibres*   97
          3.7.6   *Alumina fibre*   97
          3.7.7   *Boron fibre*   98
          3.7.8   *Metal fibres*   98

4. **High heat resistance fibres**   98
    4.1. General scope on high heat resistance fibres   98
    4.2. Inorganic fibres   99
    4.3. Thermoset type fibres   99
    4.4. Thermoplastic type fibres   99
    4.5. Fluoro-carbon fibre   99
    4.6. Flame-proof fibres   99

5. **Separation function fibres**   100
    5.1. General scope on separation function fibres   100
    5.2. Activated carbon fibre   100

	5.3.	Removal function fibre by chemical/ionic reaction	100
	5.4.	Super absorptive fibre	100
	5.5.	Membrane hollow fibres	100

6.	**Optical fibres**	**101**
	6.1.	General scope of optical fibre	101
	6.2.	Fibre for telecommunication	102
	6.3.	Fibres for lightening and for image transfer	102

| 7. | **Electric conductive fibres** | **102** |

| 8. | **Adhesive fibres** | **102** |

9.	**Dissoluble, degradable, and dissociable function fibres**	**103**
	9.1.	Dissoluble fibres	103
	9.2.	Bio-degradable fibres	103
	9.3.	Dissociable fibre	103

| 10. | **Other kinds of special function fibres** | **103** |

11.	**Specialty material fibres**	**104**
	11.1.	PVA (polyvinyl-alcohol) fibre	104
	11.2.	Polylacticacid fibre	104
	11.3.	Cellulose group fibres	104
	11.4.	Other specialty material fibres	105
		11.4.1 PEN (polyethylene naphthalate)	105
		11.4.2 PTT (polytrimethylene terephthalate)	105
		11.4.3 Metal fibres	105

12.	**Modified fibres for specific function**	**105**
	12.1.	High tenacity type	105
	12.2.	Flame retardant type	105
	12.3.	Other functionally modified types	105

13.	**Modified fibres for specific end-use**	**105**
	13.1.	Fibre-fill and cushion	105
	13.2.	Carpet	106
	13.3.	Tyre-cord	106
	13.4.	Mesh cloth	106
	13.5.	Other end-uses	106

14.	**Nano-fibres**	**106**
	14.1.	General scope on nano-fibres	106
	14.2.	Nano-fibres manufactured by bottom-up way	106
		14.2.1 Direct forming by controlled polymerization	107
		14.2.2 Gel-forming by associating monomer	107
		14.2.3 Self-organizing of polypeptide	108
		14.2.4 Self-organizing of (metal complex/organic molecule) compound	108

14.2.5   *Organic nano-tube fibre*   109
14.3. Nano-fibres produced by ESP   110
  14.3.1   *ESP processing technologies*   110
  14.3.2   *Applications of ESP*   111
14.4. Nano-fibre formed by dividing spun fibre   112
14.5. Structure, properties, and manufacturing method of **CNT**/nano-fibre   112
  14.5.1   *Structure and properties of CNT/CNF*   112
  14.5.2   *Modifications of CNT/CNF*   114
14.6. Applications of **CNT**/nano-fibre   116
  14.6.1   *Materials by homogeneous dispersion of CNT/CNF*   116
  14.6.2   *Applications for high strength fibres/threads*   116
  14.6.3   *Electronics applications of CNT/CNF*   116
14.7. Future perspectives of nano-fibres   118

15. **Concluding remarks**   118

Textile Progress
Vol. 40, No. 2, 2008, 87–121

Fibre materials for advanced technical textiles

T. Matsuo*

SCI-TEX, 12-15 Hanazono-cho, Ohtsu 520-0222, Japan

(*Received 17 April 2008; final version received 2 May 2008*)

In this article, most kinds of fibre materials used for advanced technical textiles are systematically introduced. The definition of advanced technical textiles and the scope of fibre materials used for advanced technical textiles are given in the introductory chapter, PET, nylon and PP fibres are explained as three major conventional fibres for advanced technical textiles. High mechanical performance fibres such as carbon fibre and aramid fibre, and high heat resistance fibres such as SiC fibre are introduced in chapters 3 and 4, respectively. Several kinds of function fibres such as separation function, optical, electric conductive, adhesive are introduced in chapters 5 to 10. Specialty material fibres such as PVA and PLA, modified fibres for specific function and modified fibres for specific end-use are also introduced in chapters 11 to 13. The final chapter is assigned to introduce nano-fibres which include three kinds of organic nano-fibres manufactured by bottom-up way, by electro-spinning and by top-down way, and also carbon nano-tube and nano-fibre.

Keywords: advanced technical textiles, fibre materials, high performance fibres, several kinds of function fibres, specially material fibres, nano-fibres

1. Introduction

1.1 The content and objective of this article

This article consists of 13 major chapters. In chapter 2, conventional fibres are explained, because they can quite often be fibre materials for advanced technical textile products. Chapters 3 and 4 are assigned to fibre having high performance in mechanical properties, and fibre having high performance in heat resistance. In chapters 5 to 10, several kinds of functional fibres are introduced. In chapter 11, fibres made of some specialty materials are explained. Modified fibres for specific properties and for specific end-uses are described in chapters 12 and 13. In chapter 14, several kinds of nano-sized fibres are introduced.

Objective concept for writing this book is to provide a wide view, and systematic/comprehensive knowledge source on fibre materials for advanced technical textiles, specifically for young professionals and graduated students.

1.2 What is advanced technical textiles?

It is difficult to give a strict definition for "advanced technical textiles". According to "Textile Terms and Definitions" [1], "technical textiles are: textile materials and products manufactured primarily for their technical performance and functional properties rather than their aesthetic or decorative characteristics". On the other hand, "industrial textiles" is often used in almost the same meaning as technical textiles. In this book, "technical textiles"

---

*Email: tamatsuo@nifty.com

ISSN 0040-5167 print/ISSN 1754-2278 online
DOI: 10.1080/00405160802133028
http://www.informaworld.com

is used as "textiles for non-apparel, non-household/furnishing uses. The values of technical textiles are highly based on their technical performance and functional properties". The term "advanced technical textiles" in this book means "technical textiles which have at least some technological advancement in the material and/or in the application".

1.3 Why are fibre materials used for technical products?

Generally the reason why fibre materials among several forms of materials are used for certain specified technical products is that the textile products in which some of the configurational functions of fibre are effectively utilized can have the highest value in terms of the ratio (performance/cost). The configurational functions of fibre consist of the following four elements:

(1) it is flexible (pliable),
(2) it has high ability in its axial transmission of such a physical quantity as mechanical load,
(3) it has high specific surface area, and
(4) it has technological easiness in transformability into textile structures such as weaves and non-wovens.

In apparel use, items (1), (2) especially in mechanical transmission properties, and the item (4) are generally fully used. On the other hand, in technical textiles, the item (3) often becomes the most important element.

For advanced technical textiles, the maximization in the value of the ratio (performance/cost) is pursued in their developmental process using advanced materials and/or advanced application technologies.

The performance of textiles is generally dependent on (a) material of fibre, (b) configuration of fibre and (c) assembly structure of fibre including hybrid structure with other kinds of materials/parts. It must be noted that the range in selecting these three structural elements in advanced technical textiles is by far wider than in apparel textiles.

1.4 General scope on fibre materials for advanced technical textiles

Fibre materials used for apparel textiles are quite concentrated to some specific kinds of fibres.

Cotton and polyethylene-terephthalate (PET) fibre are overwhelmingly used. Nylon fibre and acrylic fibre succeed to the above two fibres as conventional man-made fibre materials. Wool, silk, some bast fibres, some chemical fibres based on cellulose, elastane fibre, and polytrimethylene terephthalate fibre are also mentioned as minor fibre materials for apparel use. It must be noted that there are almost no other kinds of fibre than these fibre materials mentioned above. On the other hand, the aspect of fibre materials used for technical textiles is different from that for apparel textiles.

In technical textiles, PET, nylon and polypropylene (PP) are three major conventional fibre materials. But besides these conventional fibres, fibre materials for technical textiles are much diversified by their performances, functions and specialty uses. As described above, an optimization targeting for raising the value of the ratio (performance/cost) for a specific product is strongly pursued in technical textiles, which causes the selection of the optimum fibre material in terms of the ratio. The performance required is quite diversified according to the wide variety range of application products. Hence the co-existence of so many kinds of fibre materials is possible.

Table 1.   Selective List of Technical Textile Products.

Products for sustaining resource and environment	Water treatment and water production	Filtration, bio-reaction, reverse osmosis, ultra-filtration, oxidization and oil separation
	Air purification	Bag filter, air filter, removal of toxic gas, solvent recovery
	Energy resources and energy saving	Materials for battery, materials for fuel cell
	Recycle and reuse in terms of technical textiles	
Automobile	Rubber composite parts	Tyre, driving belt
	Passive safety systems	Seat belts, air bags
	Car interiors	Seat surface and door trim roof trims, floor coverings
	Filters	Engine filter, oil filter, cabin filter, filter for diesel engine gas
	Noise control sheets	
Medical/biological, hygienic	Medical/biological, para-medical	Virus removal filter, artificial kidney, artificial lunge leukocyte removal filter, tissue scaffold, DNA chip
	Hygienic	Sanitary napkin, diaper
Protection/safety	Bullet proof, stab resistance, fire fighting suit, chemical protection, radio-active protection, cold protection	
Electric and information technologies	Print circuit, electric conduction, electric insulation, optical fibre communication, acoustic uses	
Construction, civil engineering	Reinforcement for construction, geo-textiles, miscellaneous materials for construction/civil engineering	
Agriculture, horticulture, marine		
Aircraft, space	Fibre reinforced composites	
E-textiles		

1.5 General scope on technical textile products as the application items of fibre materials

In Table 1, technical textile products as the end-uses of fibre materials are selectively listed. Details of their applications to these products will be written in the successive book of advanced technical textiles series. Table 2 which shows worldwide consumption amount statistics of technical textiles reported from an investigation company [2] must be useful to have a rough idea on the market size of technical textiles by application field. There are no statistics for advanced technical textiles, which is fully dependent on the range defined as advanced technical textiles. But roughly speaking, the author feels that their share in technical textiles is over 50% in advanced countries.

Table 2. Worldwide Consumption of Technical Textiles by Application Fields [2].

	10^3 tonnes			$ million		
	2000	2005	Growth (% pa)	2000	2005	Growth (% pa)
Transport textiles (auto, train, sea, aero)	2220	2480	2.2	13080	14370	1.9
Industrial products and components	1880	2340	4.5	9290	11560	4.5
Medical and hygiene textiles	1380	1650	3.6	7820	9530	4.0
Home textiles, domestic equipment	1800	2260	4.7	7780	9680	4.5
Clothing components (thread, interlinings)	730	820	2.3	6800	7640	2.4
Agriculture, horticulture and fishing	900	1020	2.5	4260	4940	3.0
Construction – building and roofing	1030	1270	4.3	3390	4320	5.0
Packaging and containment	530	660	4.5	2320	2920	4.7
Sport and leisure (excluding apparel)	310	390	4.7	2030	2510	4.3
Geo textiles, civil engineering	400	570	7.3	1860	2660	7.4
Protective and safety clothing and textiles	160	220	6.6	1640	2230	6.3
Total above	**11340**	**13680**	**3.9**	**60270**	**72360**	**3.7**
Ecological protection textiles[a]	230	310	6.2	1270	1610	4.9

2. Conventional fibres

2.1 General scope on conventional fibres

Conventional fibres themselves can never be advanced material. But they can be converted into advanced technical textiles through advanced application technologies. As mentioned above, PET, nylon and PP can be classified as conventional fibres for technical textiles.

Properties of these three kinds of conventional fibres are summarized in Table 3 comparing p-aramid fibre as a reference. The advantages and disadvantages are summarized in Table 4.

Except in the case that high heat resistance or high specific function is needed, we must consider to select a candidate material among these conventional fibres at first; Tables 3 and 4 must be useful for the selection. Then such specification items as yarn thickness and single filament thickness, fibre length, fibre cross-sectional shape, oil used on fibre, fibre

Table 3. Properties of PET, Nylon, PP Fibres Comparing with p-Aramid Fibre (Standard Type).

Fibres	Specific gravity	Melting point (°C)	Glass transition temp(°C)	Tensile strength (MP$_a$)	Breaking elongation (%)	Tensile modulus (GP$_a$)	LOI (%)	UV degradation	Colour-change by UV
PET	1.38	260	70	510−690	15−40	6−11	18−21	*	*
nylon66	1.14	260	35(50%RH)	350−550	18−36	3.0−6.5	20−21	†	†
nylon6	1.13	220	20(50%RH)	450−700	20−32	2.5−3.4	20−21	*	†
PP	0.91	160	−15	410	25−60	6.4	18−20	‡	§
p-aramid	1.14	−	300	2760	3.3	58	29	*	‖

Note: † : fairly good, *: good, ‡: can be good by additive, §: very good by pigment colour ‖: sensitive.

Table 4.   Advantages and Disadvantages in Properties of PET and Nylon Fibres Comparing with PP Fibre.

Fibres	Advantage, feature	Disadvantage
PET	• Balanced in physical properties	• Easy in hydrolysis • A little less in wearing resistance
Nylon	• Excellent in toughness and wearing resistance	• Easy in yellowing
	• Low in modulus	• Fairy expensive as conventional fibres
	• Excellent in strain recovery	• High in energy consumption for fibre production
PP	• Excellent in chemical resistance	• Very bad in dyeability
	• Low specific gravity	• Low in heat resistance
	• Low in energy consumption for fibre production	• High in creep and stress relaxation

Abbreviations: PET, polyethylene-terephthalate; PP, polypropylene.

crimp, fibre/yarn tensile strength, breaking elongation, modulus, dimensional stability and grade of functional modification must be considered for the selection. If some ignorable insufficiency is found in the selection in terms of required performance of objective technical textile products, then we must move to select some special fibre material candidate from other kinds of fibres than these conventional fibres.

2.2 PET (polyethylene terephthalate) fibre

PET fibre is supplied in a wide variety of fibre forms among fibres such as multi-filament, textured yarn, staple fibre, mono-filament, chopped fibre and spunbond non-woven. Similarly to the case of apparel textiles, PET fibre is most popular among fibres in technical textiles. It has fairly good balance in reasonable cost and fairly good performance properties. Its mechanical properties, dimensional properties, heat resistance, light-degradation resistance and light-colour fastness are practically fairly good. But it is rather inferior in anti-hydrolysis. Disperse dyeing under high pressure for its colouring is generally adopted. But pigmented PET is available by some producers. There are several kinds of modified PET fibres also available in such function as adhesiveness, flame-retardancy, ant-pilling, anti-bacteria and soil-guard.

The fibre is most widely used among fibres for technical textiles. Tyre-cord, air-bag, car-seat, sail cloth, sewing yarn, geo-textiles, fishing net and non-woven are examples of its application products.

2.3 Nylon fibre

There are nylon 6 and 66 as major nylon fibres. The former is more economical and the latter has better temperature characteristics. They are available in the form of multi-filament, staple fibre, chopped fibre and mono-filament. But the most popular form is multi-filament.

The fibre was firstly commercialized in synthetic fibres. But its consumption amount is much less comparing to PET fibre, because it is more expensive. Further, it has lower tensile modulus, and is inferior in light fastness to PET fibre. The advantages in its properties are excellent in toughness, in strain recovery and in high flexibility. It can be easily coloured by acid dyeing.

Table 5.   Comparison of Properties among High Mechanical Performance Fibres.

Classification	Fibre material	Specific gravity	Strength (MP$_a$)	Modulus (GP$_a$)	Breaking elongation (%)
	Carbon	1.8	3600	400	1.7
	Glass	2.6	3400	78	4.0
Inorganic	Stainless steel	7.9	2400	180	1.5
	Si-C	2.4	3000	190	1.5
	Alumina	3.4	1800	300	1.0
	Boron	2.5	3600	400	0.8
	p-aramid	1.4	2900	95	3.5
Organic	PBO	1.5	5800	180	3.5
	Ultra-high MWPE	1.0	4000	95	4.0
	Polyarylate	1.4	3400	75	3.9

Its excellent toughness is utilized in tyre reinforcement. The high flexibility is useful for the compact containing in airbags. Its excellent toughness is fully utilized for floor covering in the form of bulked continuous filament.

2.4 Polypropylene fibre

Its colouring is usually conducted by blending of pigment, because it lacks dyeabilty. Such a low melting point of PP as 160°C is often insufficient for uses in technical textiles. Creep and stress relaxation of PP are also too high for some uses. These characteristics of PP fibre cause some limitation in the uses of technical textiles. On the other hand, it has advantages in the following points: most economical as material, small in energy consumption for fibre production, smallest in specific gravity as material among conventional fibres. It is highly hydrophobic and fairly feasible for waste recycling. Its commercially available fibre forms are staple fibre, multi-filament and directly spun non-woven such as spun bond and melt blown. There is also split fibre yarn produced by high uni-axial stretching of film.

PP fibre is positively used for the application products in which its disadvantages can be out-of-problem. Diaper, pad, napkin, wiper, geotextiles, needle-punch carpet, oil absorbing material, packaging, rope, tape and net are typical examples of its application products.

3. High mechanical performance fibres

3.1 General scope on high mechanical performance fibres

Properties of several high mechanical performance fibres are summarized in Table 5. Properties with additional items for organic high performance fibres by type are also summarized in Table 6.

There is a clear difference in the property pattern between inorganic and organic high performance fibres. Inorganic fibres generally have higher heat resistance and higher weather resistance than organic fibres. But they have much higher specific gravity, lower breaking elongation and are so brittle as to be easily snapped. On the other hand, organic mechanical high performance fibres are much lighter and excellent in anti-snapping. But some organic fibres such as ultra-high molecular weight polyethylene (UHMWPE) fibre have low melting temperature. The ratio (compression strength/tensile strength) of organic fibres is fairly low in contrast to inorganic fibres. They have a rather strong tendency to be easily fibrillated. Some organic fibres such as p-aramid and poly-p-phenylene benzobisoxazole

Table 6.   Comparison of Physical Properties among Organic High Mechanical Performance Fibres by their Types.

Fibre material type	Strength (GP$_a$)	Modulus (GP$_a$)	Tenacity (CN/dtex)	Specific modulus (CN/dtex)	Breaking elongation (%)	Melting or degradation temp. (°C)	Moisture regain (%)
p-aramid standard	2.9	72	20	490	3.6	550	6.5
high modulus	2.9	110	20	780	2.4	550	2.5
high tenacity	3.4	97	23	670	3.3	550	5.5
zylon AS	5.8	180	37	1150	3.5	650	2.0
(PBO) HM	5.8	270	37	1720	2.5	650	0.6
Dyneema SK60			29	990	4.0	146	0
(UHMWPE) SK71			37	1230	4.0	146	0
Vectran HT	3.4	75			3.9	330	0.2
UM	3.4	106			2.7	350	0.2

are fairly sensitive to sunlight. All the mechanical high performance fibres except glass fibre are fairly expensive.

As described above, there are no almighty fibres as high mechanical performance fibres. Therefore in practice, proper use of fibre is carried out based on the optimization principle of the ratio (performance/cost), which depends on the individual application.

3.2 Carbon fibre

There are two kinds of carbon fibre by its precursor – polyacrylonitrile and pitch. Carbon fibre is produced by carbonizing these precursor fibres under tension. The diameter of single filament is ranged from 6 to 12 μm. Carbon fibre is mostly used as reinforcing fibre for advanced composites whose major matrix is epoxy resin.

Pitch carbon fibre has lower breaking elongation, lower compression strength and is more expensive than acrylonitrile carbon fibre, but pitch carbon fibre is generally higher in modulus and in axial heat conductivity. Then it is applied only for such specified end-use as space satellite frame and precision roll. Hence its consumption amount is much less than that of acrylic carbon fibre.

The price of carbon fibre widely ranges according to its grade. Figure 1 shows a wide distribution in the mechanical properties by several grades produced by one of its leading manufacturers. The price of the fibre type having lower mechanical performance is generally less expensive.

Tow is the basic form of the fibre material whose thickness is usually expressed in the number of single filament in such a way as 12 k. Tow of standard thickness is 12 k or 24 k. Tow thicker than 48 k is called large tow, which is generally a little inferior in mechanical performance to standard grade tow of small tow, but is more economical. Carbon fibre is usually supplied in the forms of tow, prepreg tape, weave, no-crimp knitted sheet and compound for molding.

The largest advantage in the use of carbon fibre reinforced composites is weight reduction comparing to customary materials. Historically its price was so expensive that it was selectively applied to high weight sensitive end-uses such as aero-space and some special sports goods. With a technological progress for improving impact strength, they are now

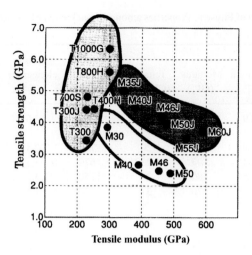

Figure 1.   Mechanical Performance by the grade of Toray carbon fibre [3].

going to be widely used for structural parts of commercial aircraft. Recently an increase in their consumption amount for industrial fields has become significant by the combination effect of technological progress in their application, lowering of fibre price and strong needs caused by social environmental problems. Hence it is forecasted that its high annual growth rate will also be kept in the future.

3.3 Glass fibre

This section focuses on glass fibre as a mechanical performance material. There are several types of glass fibre caused by its material constitution such as E-glass (non-alkali), S-glass (high strength), AR glass (alkali-resistance) and D-glass (low dielectric constant). E-glass fibre is generally used for fibre-reinforced composites. S-glass fibre is used for needs of specially high strength. AR glass fibre is mainly used for reinforcement of cement. Main use of D-glass fibre is for print circuit board. A new type of glass fibre named as Hiper-tex[TM] has recently been commercialized. It is higher by 30% in tensile strength, higher in heat-resistance and alkali-resistance than E-glass fibre [4].

Fibre diameter of glass fibre ranges mainly from 15 to 20 μm. Its cross section is usually circular. It is said that flat cross section type is more effective for reinforcement. It is very important for selecting fibre grade to have an optimum surface treatment agent in terms of good interlayer adhesiveness between fibre and matrix resin, and good process-ability. Fibre is supplied in such forms as yarn, roving, cloth, chopped strand, strand mat and milled fibre.

The advantage of glass fibre is that it has a good breaking strength and is very economical. Hence it is most widely used as reinforcing fibre for fibre reinforced plastics etc. In this sense, glass fibre can be situated as to be both high mechanical performance and conventional fibre. But it must also be noted that its specific gravity is much higher and its modulus is much lower than carbon fibre. Hence the effect of weight reduction by glass fibre reinforcement is fairly limited in comparison with carbon fibre reinforcement.

3.4 P-aramid fibre

There are some kinds of p-aramid fibres. Polyparaphenylene terephphtalate fibre is major among them, on which this section is focused. The fibre is composed of highly oriented stiff molecular. It is produced by air-gap wet spinning [5]. The positions of p-aramid fibre in mechanical performance and heat resistance among several kinds of mechanical high performance organic fibres are not so significant as shown in Table 6. But it has the longest history and the largest background of successive developmental activities for its application among them. The facts and its reasonable pricing have resulted in its leading market position of high mechanical performance organic fibre group.

Its mechanical performance and heat resistance properties (melting point), are fairly good as shown in Table 6. It also has good flame retardancy. But it must also be noted that it has slightly high moisture regain and low weather resistance among them. Similar to the other organic high mechanical performance fibres, it has fairly low value of (compression strength/tensile strength) and tendency of easy fibrillation. But this tendency is effectively utilized in the application to sliding materials and seal packing.

The fibre is supplied in the form of multi-filament, staple fibre, chopped fibre (pulp), weave/knit/braid, among which chopped fibre (pulp) is the largest in volume. Application products of p-aramid fibre are summarized in Table 7.

Table 7.   Application Examples of p-Aramid Fibre [6].

Application fields		Application examples
Rubber	Tyre	Passenger car tyre, truck tyre, racing car tyre
Reinforcement	Belt	Continuously variable transmission belt, timing belt, conveyer belt
	Hose	High pressure hose, steam hose, radiator hose
General	Rope, cable	Oil rig rope, sling rope, net rope for sport, electric cable, optical cable
Industrial use	Cord, braid	Tensioner
	Narrow weave	Sewing thread, fishing cord, tennis gut, electric cord
	Cloth	Heat resistance belt, sling belt, safety belt, tape
	Net	Membrane for construction, tent, dryer canvas, sailing cloth
	Filter	Fishing net, safety net
	Non-woven, paper	Heat resistance filter
	Geo-textile	Heat resistance felt, print circuit board, electric insulation paper
		Earth reinforcing grid, net, asphalt reinforcement, aseismatic tape,
Protection	Anti-ballistic	Bullet proof jacket, helmet
	Anti-cutting	Safety glove, safety clothing, safety shoes, safety apron, sports wear
	Anti-melting	Anti-sputtering cloth, rider suit
Asbestos	Sliding material	Brake pad, clutch facing
Replacement	Gasket	Engine gasket
	Packing	Ground packing
Cement	Building material	Curtain wall, floor material, ceiling material
Reinforcement	Civil eng. material	Concrete reinforcement, pipe
Polymer	Thermoset compos	Aircraft parts, sports goods, mechanical parts, pressure vessel
Reinforcement	FRTP	Casing, sliding parts

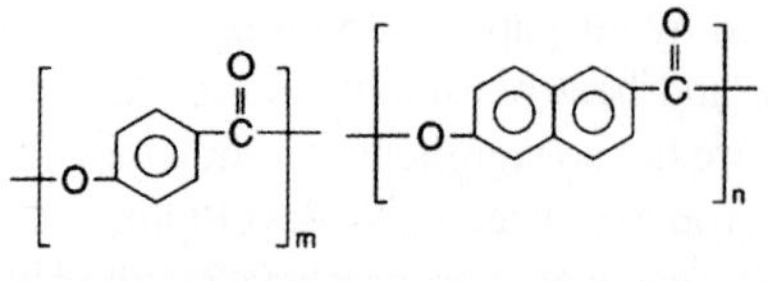

Figure 2. The chemical structure of wholly aromatic polyester fibre, Vectran™.

3.5 UHMW-PE (ultra-high molecular weight polyethylene) fibre

The fibre is composed of highly oriented crystalline in which the density of defects originated from molecular end is very small. It is usually produced by specific gel spinning and successive stretching with high draw ratio [7].

The advantages of the fibre are significantly low in specific gravity and has very high specific strength (=tenacity) as shown in Table 5. It also has high weather resistance, high impact resistance, high abrasion resistance, high flexural fatigue resistance and high vibration damping coefficient. One of its noted properties is that it has minus coefficient of linear thermal expansion. On the other hand, its melting point is as low as 146°C, its creep and stress relaxation rates are fairly large and its adhesiveness is inferior to other kinds of fibres.

Typical examples of its application products are several protective goods such as anti-ballistic clothing and helmet, rope, sport net, fishing net, fishing line, sail cloth, speaker cone and super electric conductive wire tool.

3.6 Wholly aromatic polyester fibre

The material of the fibre which belongs to wholly aromatic polyester and has been commercialized is a copolymer whose molecular structure is shown in Figure 2. It is produced by melt spinning and successive solid polymerization by curing [8].

Its mechanical properties are similar to those of p-aramid fibre. But it has lower moisture regain, better acid resistance and higher strength at low temperature, which makes it useful for space applications. But it is inferior in modulus at high temperature to p-aramid. One of its features to be noticed is that there is higher freedom of fibre making in such points as range of single filament thickness, fibre cross section, easiness in bi-component spinning and convenient use of spinning facility, because melt spinning is applicable to its production.

3.7 Other mechanical high performance fibres

3.7.1 Poly-p-phenylene benzobisoxazole fibre

The molecular structure of the poly-p-phenylene benzobisoxazole fibre is shown in Figure 3, which is rigid and straight. It has the highest mechanical strength, modulus and heat resistance among organic high performance fibres commercialized, as shown in Table 5. It also has such excellent flame retardancy as limit of oxygen index (LOI) = 70. But it is fairly sensitive to UV-light and is more expensive than p-aramid fibre.

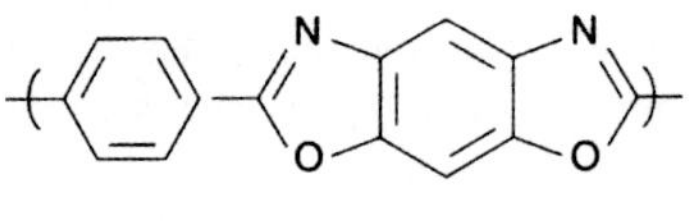

Figure 3. The chemical structure of PBO fibre.

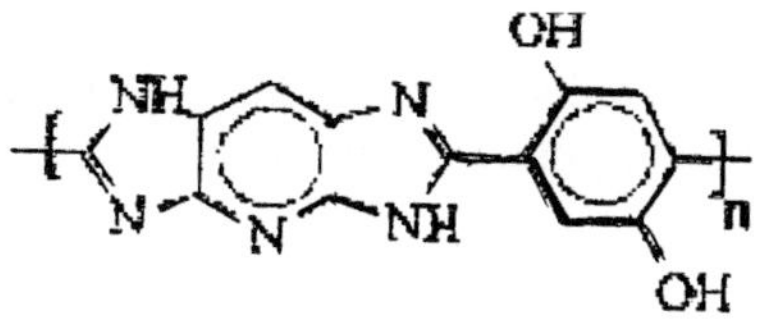

Figure 4. The chemical structure of PIPD fibre.

3.7.2 PIPD fibre

The PIPD fibre has the molecular structure as shown in Figure 4 and has almost similar mechanical performance to PBO fibre. It has the highest compression strength among organic fibres.

3.7.3 Polyketone fibre

The Polyketone fibre is a kind of copolymer; its molecular structure is shown in Figure 5. Ethylene and carbon mono-oxide are raw materials for the fibre. Hence, the cost of its raw material is very cheap. It is produced by wet spinning and successive super large ratio drawing. It has similar tenacity and modulus to p-aramid, and has higher breaking elongation (5%–7%), lower melting point (272°C) and good adhesiveness to rubber. But it is sensitive to UV light [9]. It is expected that the fibre is very useful for tyre cord.

3.7.4 Basalt fibre

The Basalt fibre is a kind of rock fibre. Its properties are slightly dependent on the local basalt rock. Its mechanical properties are almost the same as E-glass fibre. But it can be used at temperatures up to 700°C and has higher alkali-resistance than glass fibre [10].

3.7.5 SiC group fibres

The SiC group fibres are usually produced by the successive process of melt spinning of silicon polymer, in-meltable treatment in air and baking. They are composed of Si−C−O or Si−M−C−O (M = Ti or Zr). They have good mechanical properties and very good heat resistance. By excluding oxygen atoms from these molecules, the fibres obtained can become tolerable up to 2000°C [11]. They are applied to reinforcing fibre of ceramic composite materials which are useful for such an end-product as gas turbine blade.

3.7.6 Alumina fibre

Alumina fibre is defined as the fibre of polycrystalline containing alumina more than 60%. Its typical properties are shown in Table 5. But there are several kinds of alumina fibres whose properties are dependent on their production method. They are mainly used as

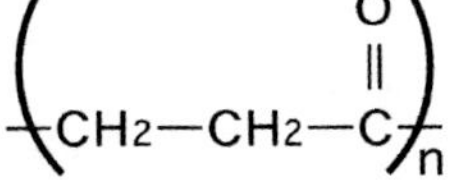

Figure 5. The chemical structure of polyketone fibre.

reinforcing material in combination with glass fibre or carbon fibre for fibre reinforced plastics and fibre reinforced metal. Its properties are much higher modulus than glass fibre, and high in compression strength, surface hardness, heat resistance and electric insulation-capability [12].

3.7.7 Boron fibre

The Boron fibre is usually produced by chemical evaporation of boron onto tungsten fibre or carbon fibre. As shown in Table 5, it has so high strength and modulus as to carbon fibre and so high compression strength as to 6.9 GPa. But its specific gravity is higher than carbon fibre. It is also very expensive. It is used for reinforcing metal which is mainly applied to specific aero-space devices [13].

3.7.8 Metal fibres

Stainless steel fibre and tungsten fibre can be mentioned as typical examples of high performance metal fibres. Stainless steel fibre has good mechanical performance in strength and modulus as shown in Table 5. But because of its high specific gravity, its specific strength and specific modulus are never good. Similarly, tungsten fibre has excellent mechanical properties in cross-sectional base. But its specific strength and modulus are not so good.

4. High heat resistance fibres

4.1 General scope on high heat resistance fibres

There are several kinds of heat resistance fibres, which are selectively used according to the requirement of several applications. Such mechanical high performance fibres as SiC fibre, alumina fibre, tungsten fibre, carbon fibre, basalt fibre, glass fibre, PBO fibre and p-aramid also have good or excellent heat resistance. But they will not be introduced again in this section. It must be noted that high heat resistance fibres also have high LOI value. Properties of typical heat resistance fibres are summarized in Table 8 for comparison. In this section, high heat resistance fibres are introduced in terms of the following classification: inorganic fibres, thermoset type fibres, thermoplastic type fibres, PTFE family fibres and flame resistance fibres.

Table 8.   Comparison in the Properties of High Heat Resistance Fibres.

Fibre material	Tenacity (CN/dTex)	Melting temp. (°C)	Degradation temp. (°C)	LOI
M-aramid	5.5		425	30
Polyamide-imide	2.7		450	32
Polyimide	3.8			38
Melamine	2.0			32
PPS	5.0	285		34
PEEK	7.2	334		33
Flame proof	2.2			55
polytetrafluoroethylene	1.6	335	400	95
Silica	35.5		1800	

4.2 Inorganic fibres

Inorganic fibres generally have much better heat resistance than organic fibres. Such fibres as alumina, silica, basalt SiC group are the typical examples. But they are so brittle as to be easily snapped. In the case of silica fibre, its softening temperature is about 1150°C and its degradation starting temperature is 1800°C. But it is much more expensive than glass fibre.

4.3 Thermoset type fibres

M-aramid fibre, poly-amid-imide fibre, poly-imide fibre and melamine fibre are typical examples belonging to this type.

M-aramid fibre is most popularly used among high heat resistance organic fibres. Examples of its application are: heat protective clothing, electric insulation material for wire and cable, interior goods of aircraft and bag filter.

Poly-imide-amid fibre is better in heat resistance, flame retardancy and chemical resistance than m-aramid fibre. Its major application products are protective clothing, working wear for chemical plants, camouflage clothing, and heat and chemical resistance filter.

Poly-imide fibre is expensive. But it is so excellent in chemical resistance at high temperature that it is mainly used for dust removal bag filter of high temperature use.

Melamine fibre is less expensive among fibres of this type and has comparatively high LOI value. Hence its major end-uses are fireman clothing, plant working wear, bed clothes, interior goods, and brake pads for automobiles.

Phenol resin fibre is less expensive among fibres of this type, and is fairly good at heat resistance, chemical resistance and flame retardancy. The burnt gas is not so toxic. Hence it is applied for fire protective goods, electric wire covering, alternatives for asbestos. It is very good precursor for activated carbon fibre.

4.4 Thermoplastic type fibres

Polyphenylene-sulphide (PPS) fibre, and polyether-ether-ketone (PEEK) fibre are typical examples belonging to this type.

Heat resistance of PPS fibre is not so good. But it is excellent in chemical resistance, anti-hydrolysis, flame retardancey and toughness. In addition, it is comparatively less expensive. Hence it is widely used for bag filter [14] of incinerators.

PEEK fibre is more expensive than PPS. But it is excellent in heat resistance, chemical resistance, anti-hydrolysis, toughness and wearing resistance. Its major applications are conveyer belt and felt used in high temperature condition [15].

4.5 Fluoro-carbon fibre

Its tensile modulus and strength is fairly low. But it has excellent heat resistance, chemical resistance, weather resistance, low friction and wearing resistance, stain resistance and high frequency electrical wave insulation. Hence its major applications are bushing for automobile suspension, air filter for semiconductor plant, bag filter, gasket, conveyer belt, cleaning blush, electrical insulation, etc [16].

4.6 Flame-proof fibres

The fibre is produced by baking organic precursor fibres such as acrylonitrile fibre in atmosphere containing oxygen. It is especially excellent in flame-proofing. Sheets for welding fire, working wear for fireplace, fire protective cloth, heat insulator and ground packing are its major application products [17].

5. Separation function fibres

5.1 General scope on separation function fibres

There are two groups of separation function fibres. In the first group, the separation takes place at the surface and/or inside the fibre. In the second group, the separation is performed by transporting material to be separated through the membrane of hollow fibres. In the following, the sub-sections from 5.2 to 5.4 are concerned with the first group and the fibres of the second group are summarized in 5.5. In the separation function fibres, the item c) "high specific surface area" (see 1.3) of the fibre configurational functions is fully utilized.

5.2 Activated carbon fibre

It can be produced by carbonizing such precursor fibre with steam activation as phenol resin fibre, cellulose fibre and acrylonitrile fibre. It has a 100 times higher adsorbing rate than granular activated carbon. It has been applied to such systems as solvent recovery system, removal system of toxic gas, deodorant system, chemical gas protective clothing, and purifier for drinking water [18].

5.3 Removal function fibre by chemical/ionic reaction

There are several kinds of chemical reactive fibres having such as photo-catalyst oxidization, adsorptive chemical reaction, absorptive chemical reaction, ionic reaction and adsorptive photo-catalyst oxidization. Their major objective functions are removal of toxic materials, removal of bacteria/virus, deodorization, and rusting guard. One of their examples is the cotton fibre within which zeolite particles are generated. Metal ion can be introduced into the zeolite. This kind of fibre is useful for anti-bacteria, anti-virus, moisture control, and water absorption [19]. Ion-exchangeable fibre is useful for the removal of harmful ionic substances within water.

5.4 Super absorptive fibre

It is produced by introducing high amounts of carboxylic acid to cross-linking polymer. Its capacity of absorbing water is as high as 30–100 times its weight. Its application examples are for water-holding in earth, stoppage of water into cable, blood absorption, dew guard, pad for incontinence, and face lotion pad [20].

5.5 Membrane hollow fibres

As major membrane hollow fibres, there are the following kinds of function: reverse osmosis, ultra-filtration/micro-filtration, dialysis and gas separation. Pore size of active layer of membrane, object to be separated, driving force for separation, membrane structure, and transportation mechanism are summarized in Figure 6 [21]. Introduction of individual fibre will be conducted in the successive book of advanced technical textile series.

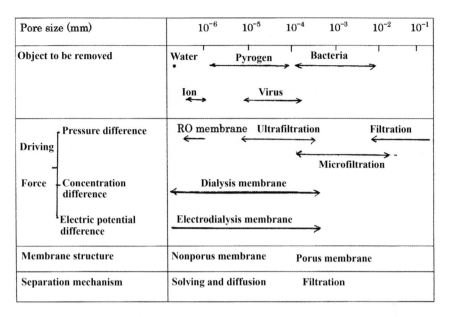

Pore size (mm)	10^{-6}	10^{-5}	10^{-4}	10^{-3}	10^{-2}	10^{-1}
Object to be removed	Water •	Pyrogen		Bacteria		
	Ion		Virus			
Driving Force — Pressure difference	RO membrane	Ultrafiltration		Filtration		
			Microfiltration			
Concentration difference		Dialysis membrane				
Electric potential difference	Electrodialysis membrane					
Membrane structure	Nonporus membrane		Porus membrane			
Separation mechanism	Solving and diffusion		Filtration			

Figure 6. The kinds of separation and its objects of membrane hollow fibres [21].

6. Optical fibres

6.1 General scope of optical fibre

Optical fibre is composed of core and clad. Light is transmitted within the core of the fibre by total internal reflection at the inner surface of clad. According to reflection mode, there are the following three types as shown in Figure 7 [22]: single mode, step index mode and graded index mode. Material of core is quart or polymer. Quart type is better in transmission loss property than polymer type (POF). But the latter can have a larger core diameter and can be more flexible than the former.

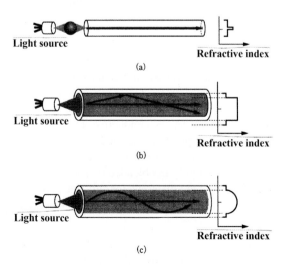

Figure 7. Three types of transmission mode in optical fibre [22].

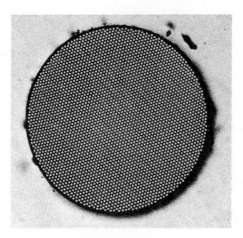

Figure 8. The cross section of image transmission optical fibre [24].

6.2 Fibre for telecommunication

For long distance telecommunication, quartz type is used, because very low transmission loss required is realized by only quartz type. To keep wide transmission band range, single mode with small core diameter is usually adopted. But for local area network, graded index mode with large core diameter is desirable, because large core is better in easiness of connection and in feasibility to low price laser than small core.

Polymethyl-methacrylate is usually used for POF because of its low transmission loss. But recently POF, whose core material is fluorinated polymer, has been commercialized, the latter has lower transmission loss than polymethyl-methacrylate POF. It is expected that POF is feasible to local area telecommunication.

6.3 Fibres for lightening and for image transfer

POF is principally used for these types of application. Rod-type fibre, which comprises of fine fibre elements of 6000 pixel (shown in Figure 8) is commercialized for image transfer use [24].

7. Electric conductive fibres

There are several kinds of electric conductive fibres in terms of its structure and production method. But they can be classified into the following three categories: (a) its whole part is made of electric conductive material, (b) its surface part is modified to be conductive and (c) some cross-sectional block part is conductive. In (a), the material can be metal, conductive polymer and polymer doped by some conductive material. In (b), there are metal platings, and are covered by conductive polymeric layer. In (c), the block part formed by bi-component spinning is made of polymer blended by some conductive material.

Major applications of the fibre are materials for anti-static electricity, electro-magnetic shielding, electric grounding, feeding of electric charge and electric circuit [25].

8. Adhesive fibres

There are two types of adhesive fibre by its structure: the fibre whose whole part is composed of adhesive material, and the bi-component fibre whose sheath part is composed of melt

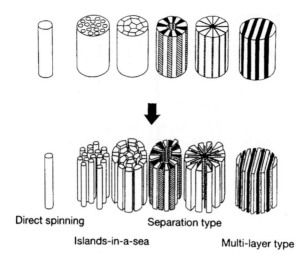

Direct spinning   Separation type

Islands-in-a-sea   Multi-layer type

Figure 9. Production principles for micro-fibres [27].

adhesive material. The latter is widely used as a binder for non-woven, paper and fibrous cushion material.

9. Dissoluble, degradable, and dissociable function fibres

9.1 Dissoluble fibres

This fibre is usually dissoluble in hot water, whose typical material is PVA. The fibre is used as a binder of paper and non-woven by its partial dissolution. It is also applied for the productions of chemical lace, and hollow yarn by its total dissolution.

9.2 Bio-degradable fibres

This fibre is easily degradable in bacteria in soil. The most popular fibre is poly-lactic-acid fibre. There is also PET fibre degradably modified by copolymerization [26]. Degradable fibre is useful especially for agricultural textiles.

9.3 Dissociable fibre

This fibre can be fibrillated or transformed into micro-fibre by mechanical and/or chemical treatment on its textile product. Its structure is usually island-sea or splittable bi-component type as shown in Figure 9 [27], or controlled phase separation blend type. It is usefully applied to man-made leather, paper, wiping cloth and filter.

10. Other kinds of special function fibres

There are some other kinds of special function fibres. Examples of these kind of fibres are (i) thermal function fibre using such materials as phase changeable agent, exo-thermal reactive agent and heat storage agent of sunlight energy; (ii) fibre having light-related function such as light reflection, light shielding and self-colouring [28] and (iii) bio-compatible function fibre for medical-use, etc.

11. Specialty material fibres

In this section, some of the fibres which do not belong to conventional fibre described in section 2 and whose function is not limited to a specific item, are described.

11.1 PVA (polyvinyl-alcohol) fibre

The fibre is conventionally produced by wet-spinning using PVA aqueous solution dope and sequential stretching for molecular orientation and crystallization. The features in its properties are comparatively high strength, hydrophilic, good alkali-resistance and good adhesiveness. Hence it has been mainly used for rope, binder for paper, reinforcement of rubber, plastic and cement.

Recently a new kind of PVA fibre which is produced by air gap wet spinning using organic solvent has been commercialized. In the new kind of fibre, there are the following three types: (a) water soluble, (b) easily fibril-able and (c) high tenacity. Main application of type (a) is binder for non-woven and paper. Type (b) fibre is mainly used for reinforcement of rubber by blend-kneading and synthetic pulp. Type (c) fibre is mainly applied to the reinforcement of cement and plastic composites [29].

11.2 Polylacticacid fibre

Its material is a kind of aliphatic polyester whose raw material is lactic-acid. It can be obtained from corn. The most important feature of the fibre is the fact that its consumption energy and carbon dioxide gas exhaust up to fibre production are lowest, respectively, among existing man-made fibres. In addition, it is usually carbon-neutral, because its low material is vegetable.

It has even better weather resistance and flame retardancy than PET. But it is gradually degraded in soil and quickly degraded in activated sludge. Its value of specific gravity, glass transition temperature, and modulus are situated about in the middle of nylon and PET. PLA fibre is conventionally made of L-stereoisomer whose melting point is 178°C. But recently, L and D stereo-complex PLA fibre whose melting point is 279°C has been commercialized [30].

Its major end-uses are as follows: (a) materials for civil engineering, (b) materials for agriculture and horticulture, (c) materials for fishery and marine, (d) materials for sanitary and medical uses and (e) materials for living disposal goods [31].

11.3 Cellulose group fibres

Natural cellulose fibres such as cotton and flax are in the minor position of technical textile field. But they will be reevaluated because of their carbon neutrality in the future. High tenacity rayon is usefully growing for high performance tyre and run flat tyre, because it has higher modulus, lower shrinkage and better adhesiveness at higher temperature than PET. Lyocell fibre, which is produced by wet spinning of cellulose using organic solvent has been developed for paper, filter and wiper by utilizing its high capability of fibril-ability. The fibre can also contain many kinds of functional agents within fibre [32]. Recently, Lyocell fibre for tyre cord has been commercialized. It has higher modulus, lower moisture regain and the same level of tenacity as rayon, compared to high tenacity rayon [33].

11.4 Other specialty material fibres

11.4.1 PEN (polyethylene naphthalate)

The fibre is the polyester in which the benzene ring of PET is replaced by naphthalene ring. Hence PEN fibre has higher by 40°C in glass transition temperature, better thermal dimensional stability and higher modulus by twice than PET. But it is more expensive than PET. It has been developed for such end-uses as high performance tyre, brake-hose and screen speaker [34].

11.4.2 PTT (polytrimethylene terephthalate)

The fibre is the polyester in which the ethylene glycol of PET is replaced by propylene glycol. Melting point and glass transition temperature of polytrimethylene terephthalate is 25°C and 20°C lower than PET, respectively. It must be noted that its modulus is lower than even nylon 6 and has excellent elastic strain recovery. The fibre is useful for stretchable fabric including non-woven and for cushion materials [35].

11.4.3 Metal fibres

Concerning metal fibres, there are several kinds of production methods such as die drawing, cutting and melt spinning. Die drawing is feasible to make thin continuous filament and also bundle composed of multi-filament. A kind of fibrous small block is produced by cutting. Concerning melt spinning, there are several ways. But the key point is how to quickly cool the molten metal during spinning. Metal fibres are applied to sintered filter, reinforcement for metal composites, electro-magnetic shielding, circuit wire and sound insulation [36].

12. Modified fibres for specific function

In this section, the fibres modified for specific function, whose base material is conventional polymer are introduced.

12.1 High tenacity type

There is, respectively, high tenacity type for PET, nylon and PP fibres. Its tenacity of nylon 6 and PP is about 10 CN/dtex, respectively.

12.2 Flame retardant type

There are flame retardant types of PET and PP fibres.

12.3 Other functionally modified types

There are several kinds of functions as modification target such as anti-bacteria, soil release, moisture absorption, wearing resistance and light weight.

13. Modified fibres for specific end-use

13.1 Fibre-fill and cushion

As typical fibre for fibre-fill and cushion, there is PET hollow fibre having 3-dimensional crimp.

13.2 Carpet

As typical fibres for carpet, there are bulked continuous nylon filament and pigmented PP stable fibre. In the former, soil release property as well as bulkiness is important.

13.3 Tyre-cord

In the case of PET fibre, high modulus and low shrinkage type is specifically used for tyre cord.

13.4 Mesh cloth

Several kinds of mono-filament such as PET, nylon, PP, PVDC (polyvinylidene chloride), PVDF (polyvinylydene fluoride), PEEK is selectively used depending on end-use requirements.

13.5 Other end-uses

As described in section 2.1, such specification items as yarn thickness and single filament thickness, fibre length, fibre cross-sectional shape, oil used on fibre, fibre crimp, fibre/yarn tensile strength, breaking elongation, modulus, dimensional stability and grade of functional modification must be optimally selected to meet the requirements of end-use. In other words, fibre manufacturer is required to provide the fibre of the grade which meets the specification of such items for each individual end-use.

14. Nano-fibres

14.1 General scope on nano-fibres

In this chapter, nano-fibre is specifically defined as the man-made fibre whose diameter ranges from 0.4 nm to 999 nm and aspect ratio is above 100. In this text, nano-tube can be also defined as the nano-fibre which is intrinsically tubular by its formation mechanism.

Electro-spinning (ESP) by which organic nano-fibre can be manufactured was invented in the 1930s. In the 1970s, DuPont Co. studied the spinning and was successful in taking high speed photography of its fibre formation. The commercial production of carbon nano-fibre (CNF) using vapor-phase growth method was started in 1988. In 1991, carbon nano-tube (CNT) was observed with its structural analyses by S. Iijima. National nano-technology initiatives presented by the US president in 2000 has made more positive effects for the activities of developing nano-technologies including nano-fibre technologies.

In the former sections of this chapter, man-made organic nano-fibres are introduced by three kinds of manufacturing methods. And then CNT/CNF is introduced. There are natural nano-fibres existing in living things. There are also nano-tube of the other materials than carbon. But they are omitted from the introduction of this chapter for page volume saving.

14.2 Nano-fibres manufactured by bottom-up way

Man-made fibres are conventionally produced by spinning bulk polymers in melt or solution. On the contrary, the objective fibres to be described in this section are produced by organizing monomers into fibres by several specific bottom-up manners. Hence, this kind of production will make the new routes fully synthesize functional fibres through molecular

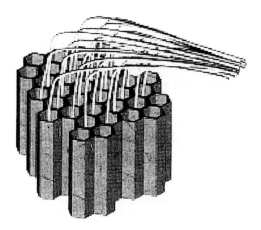

Figure 10. Schematic view of polyethylene nano-fibres growing from zeolite by direct polymerization [37].

structural design. In this sense, it can be called as highly innovative production method for
realizing extreme performance or functional fibres.

As far as the author knows, five kinds of manufacturing methods whose researches are
now at a fundamental stage can be mentioned in the followings.

14.2.1 Direct forming by controlled polymerization

Aida et al. have obtained polyethylene nano-fibre whose diameter is around 30 nm, by
polymerization utilizing catalyst contained in zeolite and introducing polyethylene gas
whose pressure is 10 atm. The fibre is composed of highly stretched chain and is as highly
crystallized as 95%. Figure 10 is a schematic view of growth of the fibres from zeolite
[37].

Masuda et al. have produced electro-conductive nano-fibre composed of poly-di-
acetylene by polymerizing self-assembled monomer by UV radiation, whose structure
is shown in Figure 11. Ookawa et al have also succeeded in electric conductive nano-circuit
polymer chain through chain reaction polymerization of di-acetylene compound by pulse
electric voltage, which is self-aligned on graphite substrate [39]. Sakaguchi has developed a
new polymerization called electro-chemical epitaxtial polymerization to obtain conductive
polymeric wire. In these process, pulse electric voltage is applied [40].

14.2.2 Gel-forming by associating monomer

Hanabusa has manufactured a series of gel fibres by associating such monomers as amino-
acid derivatives and cyclohexane-diamine derivatives [41].

Kato et al have tried to utilize this kind of nano-fibres as (a) a template for synthesizing
non-organic fibre, (b) a network structure within liquid crystal as shown in Figure 12 and (c)

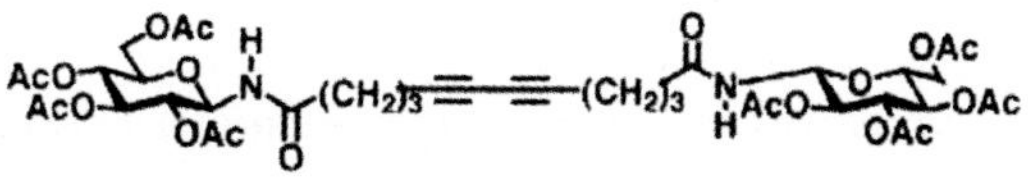

Figure 11. Dumbbell formed glycolipid derivative [38].

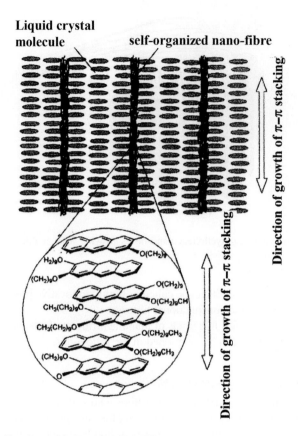

Figure 12. Nano-fibre formed in liquid crystal [42].

soft solid electrolyte for battery. In the case of (b), it is reported that nano-fibre is effective for leveling-up of the quality of liquid crystal display. In (c), such a soft solid electrolyte can keep away the leakage of electrolyte material [42].

14.2.3 Self-organizing of polypeptide

Fibre formation by self-organization of polypeptide and protein within human body causes amyloidal malady such as Alzheimer's disease. Therefore biological researches in this field have been actively carried out.

In this connection, Koga has reported nano-fibre manufactured by self-organizing polypeptide. The fibre is composed of a spirally formed molecule whose structure can be controlled by careful designing. For example, the molecule whose diameter and its length are 6 nm and a few μm, respectively, takes sinistrorse helical coil with the pitch of 50 nm. It is expected that this kind of fibre will be useful for material of bio-medical use and nano-scaled template [43].

14.2.4 Self-organizing of (metal complex/organic molecule) compound

Kimizuka has carried out study on nano-fibres made by self-organizing of (metal complex/ ionized lipid) compound. Figure 13 is one of their typical examples. In this fibre, the

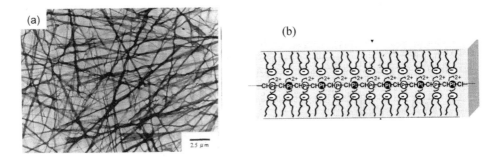

Figure 13. One-dimensional halogen bridged metal complex/anionic lipid complex fibre [44].

distance between platinum is decreased by the introduction of ionic substance. It causes an increase of electro-conductivity by non-localized charge. It is also suggested that there can be several kinds of functional fibres having such kinds of structures by combining appropriate functional molecules [44].

14.2.5 Organic nano-tube fibre

Tobacco mosaic virus is a natural organic nano-tube composed of poly-peptide. Shimizu has manufactured organic nano-tube fibres of such materials as polypeptide lipid sodium salt and glycol-lipid. In the case of the fibre shown in Figure 14, its inner diameter, can be controlled with the pitch of 1.5 nm by varying chain number n. The tube fibres can be very useful as the carrier of guest material whose size is 10–50 nm. Then it is expected that the tube fibre has a large variety of applications. It can be a nozzle for drawing a nano-width line by extruding guest fluid material which was contained in the tube by capillary effect. By solving out the tube fibre containing metal particle, metal nano-wire can be manufactured. Cd particles contained in the tube can be transformed into CdS by the reaction with H_2S. Then the tube fibres can have electro-luminescence [45].

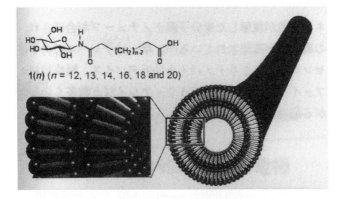

Figure 14. An example of nano-tublar fibres whose inner diameter can be varied by molecular design [45].

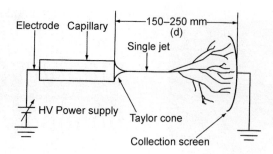

Figure 15. Illustration of a typical ESP process [46].

14.3 Nano-fibres produced by ESP

ESP is one of the most useful spinning methods to produce nano-fibre, in which fibre is formed by means of electric high voltage applied between dope supply point and electrode located behind fibre trapping part, as shown in Figure 15. ESP is usually feasible to polymer solution. Polymer liquid is thinned by electro-static traction force and is solidified by solvent evaporation during spinning.

14.3.1 ESP processing technologies

Laboratory scale apparatus of ESP is comparatively simple and less expensive. Hence ESP has been conducted in many research laboratories. But it is fairly difficult to establish the industrial technology of ESP, for which high production efficiency, product uniformity and safety are required. The problem for high production efficiency is how to economically realize large numbers of spun nano-single filaments. One way is to effectively increase the area density of nozzle hole. Yamashita has recently developed pre-industrial machines whose number of spinneret is 1000 and width is 0.6 m [47]. Another way is to make large number of dope supply points by simple apparatus without using nozzle hole. Elmalco system [48] employs the latter method. Figure 16 is the spinning observation photo of Elmarco system. Hirose paper Co. system [49] is schematically shown in Figure 17. In the system, fibres are directly spun from the dope in the bath.

The major parameters of dope related to ESP processing are viscosity (polymer concentration and molecular weight), evaporation rate of dope solvent, surface tension and electric conductivity of dope [50].

In the case that dope is polymer melt, its viscosity is usually too high to spin such a fine fibre as nano-fibre. There are two kinds of counter-measures. One is to decrease melt viscosity by applying laser to polymer at its supply point [51]. Another is to conduct spinning in vacuum or inert gas environment to enlarge the upper limit of electric spark occurrence [52].

Concerning the special form of fibre itself, technologies for making hollow, sheath/core bi-component and porous fibres have been developed. Porous fibre can be made typically utilizing phase separation between polymer and solvent. Concerning fibre assembly form, uni-axial bundle, web of controlled 3D-form and controlled pattern deposition [52, 53] are available in addition to conventional web form.

Figure 16. Elmarco system of electro-spinning [48].

14.3.2 Applications of ESP

High adaptability of polymer to ESP gives it the potentiality of a wide variety of applications. Major applications which have been commercially realized at present are air filter [54, 55] and breathable membrane [56]. ESP nano-fibre has advantage in these applications because of its very small pore size with comparatively high air permeability. Slip flow at fibre surface causes the reduction of back pressure for the filter composed of nano-fibre.

Separator for electric cell, fine transparent reinforcement, sensor, media for toxic gas removal, sustaining material for enzymes and materials for tissue engineering are typical examples of its promising applications [52, 57]. Kobayashi has tried to make artificial blood vessel of small diameter by using bio-compatible elastomer as matrix and PLA nano-fibre in the form of corrugated tube as reinforcing fibre which is obtained by ESP [53]. A Japanese non-woven manufacturer has been developing electro-spun nano-fibre sheets in pilot scale.

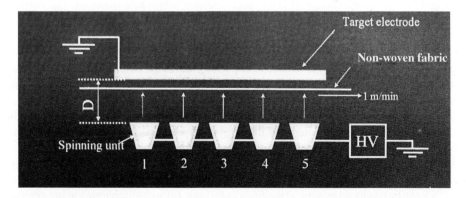

Figure 17. Hirose Paper Manufacturing Co. system of electro-spinning [49].

Figure 18. The three kinds of assembly structures in the continuous nano-fibre [59].

The major material is now PAN and the major application targets are air filter and battery separator. It is also reported that PAN nano-fibre non-woven sustaining photo-catalyst has much higher removal efficiency of VOC than the case of conventional non-woven [58].

14.4 Nano-fibre formed by dividing spun fibre

A fibre manufacturer has been successful in making the following three kinds of assembly structures of split nano-fibres, as shown in Figure 18 [59]. They are produced by solving out the sea part of stretched sea-island alloy fibres. It is also reported that nano-fibre of nylon 6 has almost the same moisture regain as cotton. Its main applications are now skin-care cloth and industrial wiping cloth [60].

Another fibre manufacturer exhibited roll samples of paper sheet made of polyester nano-fibre at ANEX 2006 [61]. It is presumed that the nano-fibre was produced through sequential application of super drawing and molecular orientation drawing of sea-island bi-component fibre tow, and then solving out the sea part.

14.5 Structure, properties, and manufacturing method of CNT/nano-fibre

14.5.1 Structure and properties of CNT/CNF

CNT is the tubular carbon whose wall is composed of carbon graphite net layer(s). The number of the layer ranges from one to dozens. Figure 19 illustrates three typical representative types of CNT classified by the different angle of the net to the tube axis. Figures 20 (a) and (b) are electro-microscopic lateral images for double wall CNT and multiple wall CNT, respectively.

Figure 21 shows the change of layer space distance versus tube diameter. The curve in the figure shows that CNT naturally becomes to CNF and then to CF with an increase in the numbers of the layers.

There are also CNT in which graphite layer is inclined to the tube axis, as shown in Figure 22, which is called as cup-stacked CNT (CSCNT). In the case of CSCNT, the surface of the tube is composed the edge of graphite layer. Hence, CSCNT has higher compatibility with matrix materials than conventional CNT.

Properties of CNT are dependent on the number of wall, and the spiral angle of graphite net to the axis. But it is estimated that the limit of its mechanical, electrical, and thermal properties are super excellent values such as tensile strength = 45 GP, current density = $1GA/cm^2$, and thermal conductivity = 6000 W/m K at room temperature. 14.5.2 Manufacturing of CNT/CNF.

CNT can be produced by several methods as follows: (a) arch-discharging using graphite electrode containing catalyst metal such as iron, (b) vaporizing of graphite containing

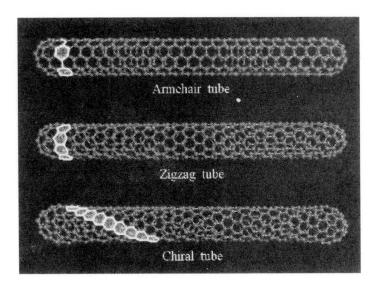

Figure 19. Representative three types of graphite structure in CNT.

catalyst metal by strong beam of laser, (c) degradation of hydrocarbon vapor mixed with fine particles of catalyst metal in the space of furnace as schematically shown in Figure 23 and (d) degradation of hydro-carbon vapor at the surface of controlled substrate sustaining metal catalyst. The method (c) is highest in production efficiency among them. The method (d) must be most excellent for controlling the structure of CNT. In any method,

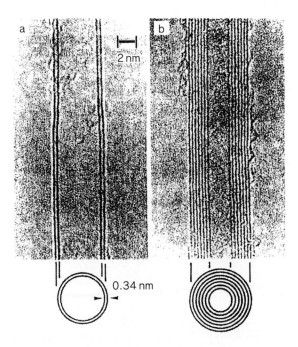

Figure 20. Electro-microscopic images of double wall CNT, and multiple wall b) CNT, respectively [62].

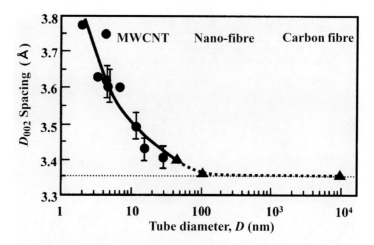

Figure 21. Change in layer spacing with tube diameter [63].

growth of CNT is carried out by metallic catalyst as suggested by the figure shown in Figure 24.

14.5.2 Modifications of CNT/CNF

CNT powder tends to so easily coagulate that it is fairly difficult to make homogeneous dispersion of CNT in matrix materials such as resin, elastomer, ceramic and metal. To realize such homogeneous dispersion and compatibility with matrix materials, several surface modifications of CNT have been tried, as shown in Figure 25. Metal plating has been tried to give good compatibility with metal matrix. Filling of iron into nano-capillary of CNT for attaining better compatibility and higher electron emissivity has been also tried.

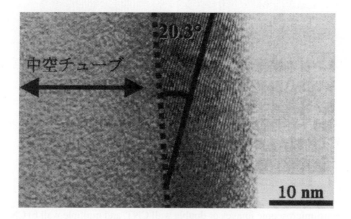

Figure 22. Lateral structure of cup-stacked CNT [63].

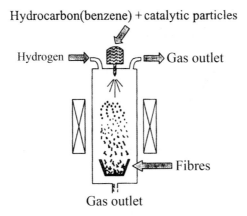

Hydrocarbon(benzene) + catalytic particles

Hydrogen ➡ ➡ Gas outlet

Fibres

Gas outlet

Figure 23. Vapor phase growth method for CNT manufacturing [63].

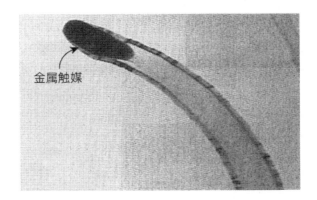

Figure 24. Metal catalyst located at the point of CNT [64].

Strategy

Aromatic Solubilizers

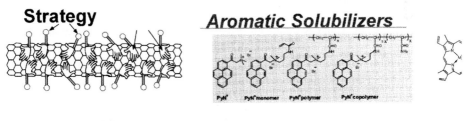

DNA & RNA

Bio-surfactants

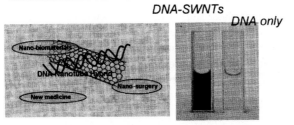

DNA-SWNTs
DNA only

X = OH, Y = COONa : Sodium cholate
X = H, Y = COONa : Sodium deoxycholate
X = OH, Y = CO-NH-(CH₂)₃-N⁺(CH₃)₂-(CH₂)₃-SO₃⁻ : Cl⁻
X = OH, Y = CO-NH-CH₂-COONa : Sodium glycocholate
X = H, Y = CO-NH-(CH₂)₂-SO₃Na : Sodium taurodeoxych

Figure 25. Examples of surface modifications of CNT [65].

14.6 Applications of CNT/nano-fibre

14.6.1 Materials by homogeneous dispersion of CNT/CNF

Film having fairly good electric conductivity with high light transmittance has been realized by homogeneous dispersion of single wall CNT [66]. By blending a few % CSCNT to matrix epoxy resin of CFRP, its impact strength can be so much increased that it is hopeful for aero-space application. Blending of CSCNT gives electric conductivity and also higher hardness to paint [67].

By blending CNT to light metal such as magnesium alloy and aluminum, their modulus and thermal conductivity can be enhanced [68].

14.6.2 Applications for high strength fibres/threads

(i) Direct thread forming from vapor phase

By introducing such hydrocarbon as methanol and some specific additive by hydrogen carrier into the furnace in which metal catalyst particles are suspended, aero-gel composed of CNT assembly is formed. It is sucked and cooled by gas flow and then wound. Thus CNT thread made from the aero-gel can be continuously obtained. Its strength is now limited to the level of 1GPa [69].

(ii) Thread spinning from CNT brush

CNT spun yarn can be manufactured by continuously pulling out and twisting a thread from CNT brush as shown in Figure 26(a). The Figure (b) is spun yarn thus obtained and braid made of the yarn. Its strength is now limited to be about 600 GPa [70].

(iii) Fibre spinning from CNT dope like liquid crystal

The dope containing single wall CNT dispersing in super acid behaves like liquid crystal in the neighborhood of 5% CNT concentration. Fibre spinning was tried from such a dope at dope temperature 100°C into such coagulation liquid as water. But the fibre thus obtained does not show any remarkable strength [71].

(iv) Fibres reinforced by CNT

There have been several trials to realize the fibres having high strength or high breaking energy by CNT reinforcement. In this section, their two remarkable examples are introduced.

By injecting single wall CNT aqueous dispersion stabilized with surfactant into the co-flowing stream of PVA aqueous solution, gel fibre was obtained. PVA fibre reinforced by CNT with the fibre volume fraction of 60% was produced by passing the gel fibre through acetone as detergent and then drying. It has the strength of 1.8 GPa and the modulus of 80 GP_a. But it must be noted that its breaking energy is 165 J/g which is much higher than 33 J/g of Kevlar [72].

PBO (see 3.3.7) was polymerized using polyphosphoric acid solution containing dispersed CNT. Then PBO fibre reinforced by CNT (5 and 10 vol %) was produced by dry-jet wet spinning [73] using coagulation bath of water and by successive annealing at 400°C. The mechanical properties of CNT reinforced PBO fibres are summarized in Table 9 in comparison with PBO fibre. By CNT reinforcement, its modulus, breaking strain, tensile strength and compressive strength is significantly increased. Dimensional stability at high temperature and creep at 400°C are also much improved by the reinforcement [74].

14.6.3 Electronics applications of CNT/CNF

(i) Applications to battery/fuel cell

CNF has been commercially used for dimensionally stabilizing the structure of Li-ion battery electrode. By its stabilization effect, the utility life of the battery has much improved, as shown in Figure 26 [75].

Table 9.   Mechanical Properties of PBO/CNT Fibres [74].

Sample	Fibre diameter (μm)	Tensile modulus (GPa)	Strain to failure (%)	Tensile strength (GPa)	Compressive strength (GPa)
PBO	22 ± 2	$138 + 20$	$2.0 + 0.2$	2.6 ± 0.3	$0.35 + 0.6$
PBO/SWNT05/5)	$25 + 2$	156 ± 20	2.3 ± 0.3	3.2 ± 0.3	0.40 ± 0.6
PBO/SWNT(95/10)	25 ± 2	167 ± 15	2.8 ± 0.3	$4.2 + 0.5$	0.50 ± 0.6

Several trials of applying CNT/CNF to fuel cell have been conducted. In an example, CNT is used as the carrier of Pt catalyst for bettering reaction efficiency at the electrode [76]. In another example, CNF non-woven is used for bettering the electric conductivity between air diffusion electrode and catalyst layer [77].

The battery whose positive electrode composed of organic radical polymer with dispersed CNF has been developed. In the electrode, CNF is effective for collecting electric charge. The battery has high power density, and high energy density. Its necessary time for

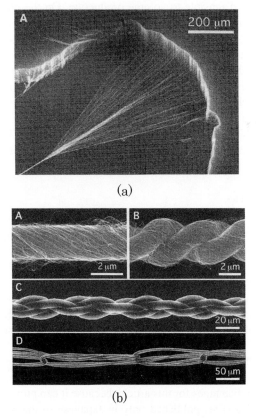

(a)

(b)

Figure 26.   Spinning process from CNT brush (a) and CNT spun yarn and braid made from the yarn (b) [70].

charging is very short. Its thickness is as thin as a flexible card [78]. Therefore it must be hopeful for power source of electronic textiles in the near future.

(ii) The other uses

It has been tried to apply CNT to electric double layer capacitor. In this application, CNT can reduce the internal electric resistance and then can realize higher out-put power. It is expected that such a capacitor can be applicable to the power source for hybrid auto-mobile [79].

CNT has been tied to apply field emission display for thin- and large-sized panels as electron emitting source. CNT can be arranged by printing on glass plate of the display [80].

Several trials of applying CNT to future LSI devices have been carried out. One is circuit wiring using CNT. Another is plates made of CNT for thermal control for multi-layered circuit substrate.

14.7 Future perspectives of nano-fibres

Concerning man-made organic nano-fibre, the fibre directly built-up by organizing monomer is most innovative among the above three kinds of manufacturing categories. But it seems to be highest in production cost. Hence, it is feasible to only extremely functional and value-added applications. Nano-fibre formed by dividing spun fibre mentioned above can be produced using conventional melt spinning system with specifically modified devices. Hence it is most feasible to volume production. On the other hand, ESP can be applicable to various kinds of materials. Therefore, it will be preferably applied to produce web of ultra-thin-layer, and nano-fibres of high value added specialty material. In these senses, these three kinds of methods for nano-fibre production will co-exist in the future.

CNF has been practically used for the electrode of Li-ion battery. But most of the application technologies of CNT/CNF is still at the stage of fundamental research. Now a huge amount of research investment in this field is being conducted worldwide. Hence it is expected that there will be some successful results for CNT/CNF in a few years. But there is also a concern if CNT has some toxic effects for human health. Therefore, it is needed to establish some standards for the handling of CNT in the very near future.

15. Concluding remarks

In the writing of this monograph, the author tried to take care on the following points: (a) it must be more convenient for the readers who want to use fibre materials for advanced technical textiles, (b) it can give a wider view to the readers by systematic introduction using general scope in the major chapters. The last chapter is assigned to nano-fibres, which are now rapidly progressing. Hence future perspectives are specially added.

The author is going to write the next monograph titled as "Advanced Technical Textile Products", which is the application part successive to this fibre part. By reading both parts, the readers will be able to attain more comprehensive knowledge on advanced technical textiles.

The author is afraid that the readers may feel some inconvenience for the literatures referred to in this article, because many of them are written only in Japanese. But this fact also means one of the advantages for this article, because it can provide lots of information on advanced technical textiles published only in Japanese to the readers who cannot so easily understand Japanese literature. To compensate the inconvenience to some extent, a few books are listed at the end of the following references.

References

[1] J.E. McIntyre et al. (eds.), *Textile Terms and Definitions*, 10th ed., The Textile Institute, Manchester, 1995, p.340.

[2] C. Byrne, *Handbook of Technical Textiles*, The Textile Institute, Woodhead Publishing Ltd., Tokyo, 2000, p.11.

[3] Pamphlet of Toray Carbon (Torayca), Toray Co. Ltd., Tokyo, 2000.

[4] Pamphlet of Hypertex™, Owen Corning World Headquarters, Toledo, Ohio, USA, 2007.

[5] H. Blades, U.S. Patent 3,767,756 (1973).

[6] O. Nakayama, J. Text. Machinery Soc. Japan 56 (2003) p.111.

[7] Y. Ohta, J. Tex. Machinery Soc. Japan 56 (2003) p.225.

[8] J. Nakagawa, J. Text. Machinery Soc. Japan 56 (2003) p.210.

[9] The pamphlet of Cyberlon™, AsahiKasei Co., Osaka, 2006.

[10] Pamphlet of technobasalt, Kyiv Ukraine, 2007. Available at http://sales@technobasalt.com.

[11] K. Okamura, SEN-I Gakkaishi, 56 (2000), p.52.

[12] H. Umesaki, *Alumina Fiber, Handbook of Fibre Materials for Industrial Uses*, The Society of Fiber Science and Technology, Japan, 1994, p.110.

[13] K. Miyasaka, *Handbook of Fibres (Sen-I Binran)*, The Society of Fiber Science and Technology, Japan, Maruzen Co., 2004, p.194.

[14] A. Wada, *Handbook of Fibre Materials for Industrial Uses*, The Society of Fiber Science and Technology, Japan, 1994, p.165.

[15] Pamphlet Zyex, Swicofil AG, Emmenbruake, Switzerland.

[16] S. Nakasuka, *Handbook of Fibre Materials for Industrial Uses*, The Society of Fiber Science and Technology, Japan, 1994, p.160.

[17] M. Hirai, *Handbook of Fibre Materials for Industrial Uses*, The Society of Fiber Science and Technology, Japan, 1994, p.173.

[18] N. Ishizaki, *Activated Carbon*, Kohdansha Co., 1992, p.189.

[19] H. Inoue, Proceedings of Review Lectures on Recent Fibre Technologies, The Fiber Society, 2007, p.7.

[20] T. Terada, *Handbook of Fibre Materials for Industrial Uses*, The Society of Fiber Science and Technology, Japan, 1994, p.317.

[21] K. Nitta et al., *Handbook of Fibre Materials for Industrial Uses*, The Society of Fiber Science and Technology, Japan, 1994, p.323.

[22] Y. Koike, *Handbook of Fibres (Sen-I Binran)*, The Society of Fiber Science and Technology, Maruzen Co., Japan, 2004, p.636.

[23] S. Takahashi, *Plastic Optical Fiber*, Koubunshi, the Society of Polymer Science, Japan, 2007, p.516.

[24] K. Shimada, *Spinning of optical fiber*, in *Advanced Fiber Spinning Technology*, T. Nakajima, ed., Woodhead Publishing Ltd, Cambridge, UK, 1994, p.222.

[25] H. Aaie, Fiber Preprints, vol 62, No 1, The Society of Fiber Science and Technology, Japan, 2007, p.13.

[26] Biomax, the home page of DuPont, 2007.

[27] M. Okamoto, *Spinning of ultra-fine fibers*, in *Adavanced Fiber Spinning Technology*, T. Nakajima, ed., Woodhead Publishing Ltd, Cambridge, UK, 1994, p.189.

[28] Pamphlet of Morphotex™, Teijin Fiber Co. Ltd., 2004.

[29] A. Oomori, *Handbook of Fibres (Sen-I Binran)*, The Society of Fiber Science and Technology, ed., Maruzen Co., Japan, 2004, p.176.

[30] K. Toyohara et al., Fiber Preprints, Japan, Autumn Meeting, 2007. vol 62, No 3. p.112.

[31] M. Mochizuki, *Handbook of Fibres (Sen-I Binran)*, The Society of Fiber Science and Technology, ed., Maruzen Co., Japan, 2004, p.878.

[32] T. Shulze et al., *AluSERU-A Versatile Material Shaping Technology*, paper presented at Fiber Society 2005, Spring Conference, St. Gallen, Switzerland, 2005.

[33] T.-J. Lee, Fiber Preprints, Annual Meeting, Japan, 2007, vol. 62, No 1, p.102.

[34] H. Mori, *Properties and Applications of Polyethylene–Nophthalene Fiber*, Extended Abstract of 10th Japan Sea Polymer Workshop, the Society of Polymer Science, Hukui, Japan, 2006, p.19.

[35] T. Matsuo, *Handbook of Fibres (Sen-I Binran)*, The Society of Fiber Science and Technology, ed., Maruzen Co., Japan, 2004, p.165.

[36] K. Miyasaka, *Handbook of Fibres (Sen-I Binran)*, The Society of Fiber Science and Technology, Maruzen Co., ed., Japan, 2004, p.194.

[37] T. Aida et al., High Polym. Jpn. 50 (2001) p.766.

[38] M. Masuda et al., 2007. Available at http://www.aist.go.jp/NIMC/recent/r99-06-01.htm.

[39] Y. Ookawa et al., Koubunnshi 56 (2007) p.436.

[40] H. Sakaguchi, Koubunshi 56 (2007) p.435.

[41] K. Hanabusa, Sen-I Gakkaishi 58 (2002) p.213.

[42] T. Katoh et al., Sen-I Gakkaishi 59 (2003) p.18.

[43] T. Koga, High Polym. Jpn. 55 (2006) p.154.

[44] N. Kimizuka, High Polym. Jpn. 55 (2006), p.138.

[45] T. Shimizu, Lecture text, 8th nano-fibre technology forum, Society of Fiber Science and Technology, Japan, 2006, p.21.

[46] A. Buer et al., Text. Res. J. 71 (2001) p.323.

[47] Y. Yamashita: Kakougijutsu 42 (2007) p.346.

[48] Pamphlet of Elmarco Co. Ltd.: at ANEX 2005.

[49] Y. Kishimoto, Slides of PowerPoint presented at symposium held by Textile Machinery Society, Japan, 2007.

[50] Y. Yamashita, Kakougijutu 41 (2006) p.459.

[51] N. Ogata, Preprint of Autumn Conference, Society of Fiber Science and Technology, Japan, Vol. 61, 2006, p.38.

[52] A. Greiner et al., Proceedings of International Nanofiber Symposium, Vol. 39, 2007.

[53] N. Kobayashi et al., J. Text. Machinery Soc. Japan 59 (2006) p.325.

[54] T.H. Grafe, Proceedings of 1st International Congress on Nano-fiber Science and Technology, Vol. 44, 2004.

[55] J.C. Binzer, Proceedings of 1st International Congress on Nano-fiber Science and Technology, Vol. 54, 2004.

[56] J. George, Proceedings of Avantex Symposium, AX.2.3, 2007.

[57] M.M. Hussain, Ind. J. Text. Res. 31 (2006) p.41.

[58] M. Kawabe, Kakou-gijutu 41 (2006) p.477.

[59] Pamphlet of Toray. Co., at Nano-Tech 2006 Exhibition, 2006.

[60] T. Ochi et al., The Investigation of Nanofibers by Mult Spinning, SEN-I-Gakkaishi, The Society of Fiber Science and Technology, Japan, 63, 2007, p.423.

[61] Exhibited products at Teijin Co. Stand, at ANEX, 2006.

[62] S. Iijima, Proceedings of Advanced Technology Forum, 2001, p.41.

[63] M. Endoh, Sen-I Gakkaishi 59 (2003) p.413.

[64] B. Maruyama, SAMPE J. 38 (2002) p.59.

[65] The Pamphlet of Soluble Carbon Nanotube, Nagasaki University, Nakashima Lab., NEDO at Nano-tech Exhibition, 2006.

[66] The Pamphlet of Technology for Dispersion of Carbon Nanotubes, AIST/MITSUBISHI Rayon Co., NEDO, at Nano-tech Exhibition, 2006.

[67] N. Ishiwatari, Proceedings of SAMPE JISTES 2006 Kyoto, 2006.

[68] A pamphlet of NEDO at Nano-tech Exihibition 2006.

[69] Y. Li et al., Sci. Exp. 11 March 1 (2006).

[70] M. Zhang et al., Science 306 (2004) p.1358.

[71] L.M. Ericson et al., Science 305 (2004) p.1447.

[72] A.B. Dalton et al., Nature 423 (2003) p.703.

[73] H.H. Yang et al., *Fiber spinning of anisotropic polymers*, in *Advanced Fiber Spinning Technology*, Woodhead Publishing Ltd., Cambridge, UK, 1994, p.142.

[74] S. Kumar et al., Macromolecules 35 (2002) p.9093.

[75] M. Endoh, Extended abstract of Summer Seminar held by Sen-I Gakkai (2002) p.75.

[76] The pamphlet of Fuel Cells Using Carbon Nanohorns (CNHs), NEC Sorp., NEDO at Nano-tech Exihibition, 2005.

[77] T. Koyama, Extended abstract of 8th Nano-fiber Technological Strategy Meeting held by Sen-I Gakkai (2006) p.31.

[78] Y. Kubo, Extended abstract of 8th Nano-fiber Technological Strategy Meeting held by Sen-I Gakkai (2006) p.4.

[79] The pamphlet of Double Layered Capacitor, Ueda Local Intelligent Cluster, for Business Development at Nano-tech Exihibition 2005.

[80] M. Okai, Extended abstract of 8th Nano-fiber Technological Strategy Meeting held by Sen-I
      Gakkai (2006) p.13.

Bibliography published by Woodhead

General

A. R. Horrocks et al., eds., *Handbook of Technical Textiles*, 2000.
YEL-Mogahzy. *Fiber to Fabric Engineering, From Traditional to Technical Textiles*, 2006.
N. Pang et al., *Fluid transport phenomena in fibrous materials,* Text. Prog. 38, No2 (2006).

Fibre materials

T. Hongu et al., *New Millennium Fibres* 2005.
J. W. S. Hearle et al., *Physical Properties of Textile Fibres*, 4th ed. 2008.
B. L. Deopura et al., eds., *Polyesters and Polyamides*.2008.
J. W. S. Hearle, eds., *High Performance Fibres*, 2001.
J. W. S. Hearle et al, *Physical Properties of Textile Fibres*, 4th ed., 2008.
B. L. Deopura et al, eds., *Polyesters and Polyamides,* 2008.
J. E. McIntyre, ed., *Synthetic Fibres: Nylon, Polyester, Acrylic, Polyolefin*, 2004.
J. G. Cook, *Handbook of Textile Fibres*, 2nd ed., 1984.
S. Adanur, *Wellington Sears Handbook of Industrial Textiles*, 1995.